Mastering

iOS 18

Is Your iPhone Ready for
the Future of Smart
Technology?

Joseph G. Seng

Table of Contents

Chapter 1: Introduction to iOS 18

Overview of iOS 18

Apple's iOS 18 represents the latest leap forward in the evolution of its mobile operating system, bringing a host of innovative features designed to enhance user experience, increase productivity, and seamlessly integrate with the broader Apple ecosystem. With each iteration, iOS has pushed the boundaries of what mobile devices can achieve, and iOS 18 is no exception. This chapter delves into the new and exciting features of iOS 18, its historical context, the importance of regular software updates, and how to prepare for upgrading to this cutting-edge operating system.

New Features and Enhancements

Apple Intelligence: At the heart of iOS 18 is Apple Intelligence, a sophisticated system that leverages personal context to provide intuitive, relevant assistance tailored to the user's needs. This feature enhances Siri's capabilities, making the virtual assistant more context-aware and responsive than ever. Users can expect more accurate suggestions, personalized content creation, and improved task automation.

Customization Options: iOS 18 offers unprecedented customization possibilities. Users can now rearrange apps and widgets more freely, customize app icons with new dark themes and tint options, and protect sensitive apps with locking and hiding features. The redesigned Control Center allows for a more fluid user experience, providing quick access to frequently used controls and the ability to personalize the layout.

Photos App: The Photos app has undergone a significant redesign, featuring automatic organization by topics such as Recent Days, Trips, and People & Pets. The new Carousel mode highlights a curated selection of photos each day, presenting them in a visually appealing, poster-like format.

Messages: Communication is elevated with new text effects, animated enhancements, and the ability to send messages via satellite, ensuring connectivity even in remote areas. Users can now schedule messages with the Send Later feature and enjoy richer media experiences with support for RCS (Rich Communication Services) messaging.

Safari and Security: Safari introduces Highlights, which automatically detect and emphasize relevant information on web pages, along with a redesigned Reader mode that provides summaries and tables of contents for articles. The new Passwords app centralizes credential management and synchronization across devices, enhancing security and ease of access.

Gaming and Entertainment: Game Mode optimizes performance for prolonged gameplay by minimizing background activities and reducing audio latency for AirPods. Personalized Spatial Audio transforms the gaming experience, immersing users in their favorite games.

Productivity and Well-being: The Notes app now supports live audio transcription and quick calculations, while the new Journal app helps users log their state of mind, track goals, and integrate entries with the Health app.

Privacy and Security: Advanced privacy features include improved control over contact sharing, enhanced Bluetooth privacy, and a redesigned Privacy and Security settings menu.

Compatibility and Availability

iOS 18 is compatible with a wide range of devices, including iPhone 15, 14, 13, 12, 11, XS, XR, and SE (2nd generation and later). This broad compatibility ensures that a significant portion of Apple's user base can benefit from the new features and enhancements. The public release is expected in the fall, with a beta version available for early adopters.

History and Evolution of iOS

The journey of iOS from its inception to its current state is a fascinating story of technological innovation, market adaptation, and relentless improvement. iOS, initially introduced in 2007 with the launch of the first iPhone, has continually evolved to meet the growing demands of users and stay ahead of technological advancements.

Early Beginnings: iOS 1 to iOS 3

The original iPhone OS, later renamed iOS, was a revolutionary departure from the mobile operating systems of its time. It introduced a multi-touch interface, a virtual keyboard, and the App Store, which allowed developers to create and distribute applications. The early versions of iOS laid the groundwork for many of the features we take for granted today, such as web browsing, email, and multimedia capabilities.

iOS 1 featured a simple, intuitive interface with basic apps like Safari, Mail, and iPod. **iOS 2** introduced the App Store, opening the floodgates for third-party app development and transforming the iPhone into a versatile platform. **iOS 3** added functionality such as cut, copy, and paste, MMS, and spotlight search, enhancing the usability of the system.

Expansion and Innovation: iOS 4 to iOS 7

With the release of **iOS 4**, Apple introduced multitasking, a unified inbox, and the Game Center, marking a significant leap in functionality. The introduction of the Retina Display in **iOS 4** also set new standards for screen resolution and visual clarity.

iOS 5 brought iCloud integration, Notification Center, and iMessage, enhancing connectivity and synchronization across devices. **iOS 6** replaced Google Maps with Apple Maps and introduced Siri, Apple's intelligent assistant, which has since become a cornerstone of the iOS experience.

iOS 7 represented a major visual overhaul, moving away from skeuomorphic design to a flatter, more modern aesthetic. This version also introduced Control Center and AirDrop, making it easier for users to access controls and share content.

Maturity and Refinement: iOS 8 to iOS 11

iOS 8 built upon the foundation of its predecessors by introducing Continuity, which allowed seamless integration between iOS and macOS devices, and HealthKit, which paved the way for health and fitness tracking.

iOS 9 focused on performance improvements and introduced features like Proactive Assistant and Transit Directions in Maps. **iOS 10** brought a redesigned Lock Screen, rich notifications, and an overhauled Messages app with stickers and animations.

iOS 11 continued the trend of refinement with features such as a new Files app, improved multitasking for iPad, and augmented reality capabilities through ARKit.

Recent Developments: iOS 12 to iOS 17

iOS 12 emphasized performance improvements, particularly for older devices, and introduced Screen Time to help users manage their device usage. **iOS 13** brought a system-wide dark mode, a revamped Photos app, and significant improvements to privacy controls.

iOS 14 revolutionized the Home Screen with the introduction of widgets and the App Library, offering a more customizable and organized user experience. **iOS 15** focused on enhancing FaceTime with spatial audio and SharePlay, improving notifications, and adding Focus modes for managing distractions.

iOS 16 introduced major updates like Lock Screen customization, Live Activities, and significant improvements to Messages and Mail. **iOS 17** further refined the user experience with enhanced sharing options, Journal app for mental health, and improved performance across the board.

The Significance of iOS 18

iOS 18 represents the latest chapter in this evolution, bringing together the best features of its predecessors while introducing groundbreaking new capabilities. It builds on Apple's legacy of innovation, integrating artificial intelligence, enhanced customization, and deeper ecosystem integration to offer a truly transformative experience.

The Importance of Software Updates

Regular software updates are crucial for maintaining the security, functionality, and performance of any operating system. For iOS, updates provide a range of benefits that extend beyond simply adding new features. They address security vulnerabilities, improve system stability, enhance compatibility with new apps and devices, and ensure a consistent and optimal user experience.

Security Enhancements

One of the primary reasons for regular software updates is to address security vulnerabilities. As technology evolves, so do the methods employed by malicious actors to exploit system weaknesses. Apple consistently updates iOS to patch these vulnerabilities, protect user data, and safeguard against potential threats. Each new version of iOS includes security enhancements designed to counteract emerging threats, making regular updates essential for maintaining device security.

Performance Improvements

Software updates also play a vital role in optimizing system performance. Over time, older versions of an operating system may become less efficient due to accumulated updates, app compatibility issues, and evolving hardware. Regular updates ensure that the operating system runs smoothly, taking full advantage of the latest hardware capabilities and software optimizations.

Compatibility and New Features

As new apps and hardware devices are released, compatibility issues can arise if the operating system is not kept up to date. Regular updates ensure that iOS remains compatible with the latest apps and devices, providing users with access to new features and improved functionality. This is particularly important in a rapidly evolving technological landscape where new innovations are introduced regularly.

User Experience Enhancements

Beyond security and performance, regular updates contribute to a better user experience by introducing new features and enhancements. These updates can include improvements to existing functionalities, new app integrations, and refinements to the user interface. By keeping their devices up to date, users can enjoy the latest innovations and make the most of their devices.

Addressing Bugs and Issues

No software is perfect, and updates often include fixes for bugs and issues that have been identified in previous versions. These fixes improve the overall stability and reliability of the operating system, reducing the likelihood of crashes, glitches, and other problems that can affect the user experience.

Examples from iOS History

Throughout the history of iOS, updates have consistently brought significant improvements and innovations. For instance, the introduction of iOS 5 brought iCloud and iMessage, revolutionizing how users synchronize data and communicate. iOS 7's major visual redesign modernized the user interface, while iOS 12's performance enhancements made older devices more efficient. Each of these updates illustrates the importance of regular software updates in keeping the operating system current, secure, and functional.

Preparing for the Upgrade

Upgrading to a new version of iOS, such as iOS 18, involves several important steps to ensure a smooth transition and minimize potential issues. Proper preparation helps users take full advantage of the new features and enhancements while safeguarding their data and minimizing disruptions.

Backup Your Data

Before upgrading, it is essential to back up all important data. This can be done through iCloud or iTunes, providing a safety net in case anything goes wrong during the upgrade process. Backing up data ensures that photos, contacts, app data, and settings can be restored if needed.

Check Device Compatibility

Not all devices are compatible with every iOS update. Users should verify that their device is eligible for iOS 18 by checking Apple's official compatibility list. Ensuring compatibility helps prevent potential issues and guarantees that the device can support the new features and enhancements.

Free Up Storage Space

Installing a new version of iOS typically requires a significant amount of storage space. Users should free up space by deleting unnecessary apps, files, and media. This helps facilitate a smooth installation process and ensures that there is enough room for the new update.

Update Apps

It is advisable to update all installed apps to their latest versions before upgrading to a new iOS. App updates often include compatibility fixes and optimizations for the latest operating system, reducing the likelihood of app-related issues after the upgrade.

Ensure Adequate Battery Life

Upgrading to a new iOS version can be a time-consuming process, especially if the update is large. Users should ensure that their device has sufficient battery life or is connected to a power source during the upgrade to avoid interruptions.

Be Prepared for Possible Issues

While iOS upgrades are designed to be as seamless as possible, there can sometimes be unforeseen issues. Users should be prepared for potential minor glitches or compatibility problems and should follow Apple's troubleshooting guidelines if any issues arise.

Follow Official Upgrade Instructions

Apple provides detailed instructions for upgrading to the latest iOS version. Users should follow these instructions carefully to ensure a successful installation. This includes downloading the update through the Settings app and following the on-screen prompts.

Post-Upgrade Steps

After upgrading to iOS 18, users should take some time to explore the new features and settings. Familiarizing themselves with the changes will help them make the most of the new capabilities and ensure that their device is configured to their preferences.

iOS 18 represents a significant milestone in the ongoing evolution of Apple's mobile operating system, offering a range of innovative features and enhancements designed to improve the user experience. Understanding the history and evolution of iOS, the importance of regular software updates, and how to prepare for an upgrade are essential for users who want to make the most of this latest release. By embracing iOS 18, users can enjoy a more personalized, efficient, and secure mobile experience, fully leveraging the power of their devices in the modern digital landscape.

Chapter 2: Apple Intelligence

What is Apple Intelligence?

Apple Intelligence marks a transformative leap in the realm of digital assistance and user interaction. At its core, Apple Intelligence leverages advanced machine learning algorithms and contextual analysis to deliver a personalized and intuitive user experience. This innovative system is embedded within iOS 18, redefining how users interact with their devices by offering real-time assistance, personalized recommendations, and proactive features tailored to individual needs.

Key Components of Apple Intelligence:

- **Contextual Awareness:** By understanding the user's habits, preferences, and current activities, Apple Intelligence provides timely suggestions and relevant information. This context-aware approach allows the system to anticipate user needs, streamline tasks, and enhance productivity.
- **Integration with Siri:** Siri, Apple's virtual assistant, is now more powerful and versatile, thanks to its deep integration with Apple Intelligence. This integration enables Siri to provide richer responses, understand

nuanced queries, and offer contextually relevant assistance across various applications.

- **Enhanced Machine Learning:** Apple Intelligence utilizes advanced machine learning models to analyze user behavior, preferences, and interactions. These models are continually updated, ensuring that the system adapts and evolves to meet the changing needs of users.
- **Personalized Recommendations:** Apple Intelligence offers personalized recommendations across various domains, including apps, content, and system settings. These recommendations are based on the user's historical usage patterns and current context, providing a tailored experience that enhances usability and convenience.

Personal Context and Its Benefits

Personal context is the cornerstone of Apple Intelligence, enabling the system to deliver a more intuitive and personalized user experience. By leveraging data such as location, time, app usage, and user preferences, Apple Intelligence can anticipate user needs and provide relevant suggestions.

Benefits of Personal Context:

- **Proactive Assistance:** Apple Intelligence uses personal context to offer proactive assistance, such as suggesting apps or actions based on the user's current activity or location. For example, if a user frequently orders food from a particular restaurant at lunchtime, Apple Intelligence can suggest the restaurant and display the menu when lunchtime approaches.
- **Enhanced Productivity:** By understanding the user's daily routine and preferences, Apple Intelligence can streamline tasks and reduce the need for manual input. For instance, it can automatically prioritize notifications, suggest relevant documents for upcoming meetings, or remind users of important deadlines based on their calendar entries.
- **Seamless Integration:** Personal context allows Apple Intelligence to integrate seamlessly with other Apple services and apps, creating a cohesive and unified user experience. This integration extends to third-party apps, enabling them to utilize Apple Intelligence's contextual capabilities for enhanced functionality.
- **Improved User Experience:** The ability to anticipate user needs and provide timely suggestions enhances the overall user

experience, making interactions with the device more fluid and intuitive. Users can enjoy a more personalized and efficient experience, with less effort required to complete tasks.

Enhanced Writing Tools

iOS 18 introduces a suite of enhanced writing tools powered by Apple Intelligence, designed to assist users in creating, summarizing, and organizing written content more effectively. These tools leverage advanced natural language processing (NLP) capabilities to provide real-time suggestions and improvements.

Writing and Summarizing

- **Intelligent Suggestions:** Apple Intelligence offers intelligent writing suggestions, helping users enhance their writing style, grammar, and clarity. These suggestions are contextually relevant and tailored to the user's writing habits, making the writing process more efficient and enjoyable.
- **Automatic Summarization:** The system can automatically summarize lengthy texts, extracting key points and providing concise overviews. This feature is particularly useful

for reviewing documents, articles, or emails, allowing users to quickly grasp the main ideas without reading the entire text.
- **Contextual Enhancements:** Apple Intelligence can analyze the context of the written content and provide enhancements such as synonym suggestions, sentence restructuring, and tone adjustments. These enhancements help users convey their message more effectively and adapt their writing to different audiences or purposes.

Notification Prioritization

- **Context-Based Prioritization:** Apple Intelligence prioritizes notifications based on the user's current context, such as time of day, location, and ongoing activities. Important notifications are highlighted, while less critical ones are deferred or grouped together, reducing distractions and helping users focus on what matters most.
- **Adaptive Notification Management:** The system adapts to the user's preferences and behavior, learning which notifications are most relevant and adjusting prioritization accordingly. This adaptive approach ensures that users receive timely and relevant information without being overwhelmed by unnecessary alerts.

- **Customizable Settings:** Users can customize notification prioritization settings to align with their preferences and needs. This customization includes options for adjusting priority levels, setting quiet hours, and choosing which apps can bypass the prioritization system for critical alerts.

Image Creation and Genmoji

Apple Intelligence extends its capabilities to the realm of visual content creation, offering tools for generating images and creating personalized emojis, known as Genmoji.

Image Creation

- **AI-Powered Image Generation:** Apple Intelligence uses advanced AI algorithms to generate high-quality images based on user input. Users can describe the desired image, and the system creates a visual representation that matches the description. This feature is useful for creating custom graphics, illustrations, and visual content for various purposes.
- **Customization Options:** Users can customize generated images by adjusting parameters such as color, style, and composition. Apple Intelligence provides a range of options for fine-tuning the image to

meet specific requirements, ensuring that the final result aligns with the user's vision.

- **Seamless Integration:** Generated images can be easily integrated into other apps and services, such as Notes, Messages, and social media platforms. This integration streamlines the process of creating and sharing visual content, making it accessible and convenient for users.

Genmoji

- **Personalized Emojis:** Genmoji allows users to create personalized emojis that reflect their appearance, style, and preferences. Users can customize facial features, hairstyles, accessories, and expressions to create unique and expressive emojis that represent their identity.
- **Contextual Usage:** Genmoji can be used across various apps and services, including Messages, social media, and email. The system offers suggestions for using Genmoji based on the context of the conversation, enhancing communication and adding a personal touch to messages.
- **Dynamic Updates:** Apple Intelligence continually updates Genmoji options based on user feedback and trends, ensuring that users have access to the latest styles and expressions. This dynamic approach keeps

Genmoji fresh and relevant, reflecting evolving user preferences and cultural trends.

The New Era of Siri

Siri, Apple's iconic virtual assistant, has been significantly enhanced in iOS 18, marking the beginning of a new era of intelligent and contextual assistance. Apple Intelligence powers these enhancements, making Siri more capable, responsive, and contextually aware than ever before.

Richer Language Understanding

- **Natural Language Processing:** Siri's language understanding capabilities have been greatly improved through advanced natural language processing techniques. Siri can now comprehend and respond to more complex and nuanced queries, providing accurate and relevant answers to a broader range of questions.
- **Multilingual Support:** Siri supports a wider array of languages and dialects, allowing users from diverse linguistic backgrounds to interact with the assistant in their preferred language. This expanded support enhances accessibility and usability for a global audience.

- **Contextual Responses:** Siri can provide contextually relevant responses based on the user's current activity, location, and preferences. For example, Siri can suggest nearby restaurants when asked for dining recommendations or provide weather updates tailored to the user's travel plans.
- **Conversational Flow:** Siri's conversational abilities have been refined to support more natural and dynamic interactions. Users can engage in multi-turn conversations, where Siri remembers previous queries and provides coherent and contextually appropriate follow-up responses.

Contextual Awareness

- **Deep Integration with Apple Intelligence:** Siri's integration with Apple Intelligence enables a higher degree of contextual awareness, allowing the assistant to offer personalized assistance based on the user's habits and preferences. This deep integration enhances Siri's ability to anticipate user needs and provide timely and relevant suggestions.
- **Task Automation:** Siri can automate routine tasks based on contextual cues, such as setting reminders, creating calendar events, or sending messages. Users can specify triggers and conditions for these

automations, making Siri a powerful tool for managing daily activities and workflows.

- **Proactive Assistance:** Siri proactively offers assistance by analyzing the user's context and predicting potential needs. For example, Siri might suggest contacting a frequently called person during specific times or remind the user to complete tasks based on their calendar entries.
- **Integration with Third-Party Apps:** Siri's contextual capabilities extend to third-party apps, allowing developers to integrate Siri's assistance into their applications. This integration enhances the functionality of third-party apps and provides users with a seamless and cohesive experience across different platforms.

Apple Intelligence represents a significant advancement in the capabilities of digital assistance, bringing a new level of personalization, context-awareness, and functionality to iOS 18. By understanding personal context, enhancing writing tools, enabling creative image generation, and transforming Siri into a more intelligent and responsive assistant, Apple Intelligence offers a comprehensive suite of features designed to enhance user experience, productivity, and creativity. This chapter has explored the key components and benefits of Apple Intelligence, setting the stage for a deeper exploration of how these innovations are reshaping the way users interact with their devices and the broader Apple ecosystem.

Chapter 3: Customization

In the age of personalization, iOS 18 takes user customization to new heights, empowering individuals to tailor their devices to reflect their unique preferences and lifestyles. This chapter delves into the various customization options available in iOS 18, demonstrating how users can create a personalized and intuitive digital environment.

Personalizing Your Home Screen

The home screen is the heart of the user experience on any mobile device, and iOS 18 introduces a range of features to make it truly your own.

Rearranging Apps and Widgets

The ability to rearrange apps and widgets allows users to organize their home screens for efficiency and aesthetic appeal. This functionality is not just about placing icons; it's about creating a personalized workspace that enhances productivity and visual enjoyment.

- **App Placement:** Users can drag and drop app icons to their preferred positions on the home screen. This feature is useful for grouping related apps together, such as placing social media apps in one area and productivity tools in another. Additionally, users can move apps to the App Library to declutter their main screens, keeping only the most frequently used apps in easy reach.

- **Widget Customization:** Widgets provide at-a-glance information from apps and are now more flexible than ever. In iOS 18, widgets come in various sizes and can be placed anywhere on the home screen. Users can choose from a range of widget options, such as weather updates, calendar events, or health metrics, and position them alongside app icons to create a dynamic and informative layout.

- **Smart Stacks:** iOS 18 introduces Smart Stacks, which are collections of widgets that dynamically change based on the user's habits and the time of day. This feature allows users to make the most of their home screen real estate by displaying relevant information at the right time, such as showing the news in the morning and fitness stats in the afternoon.

Customizing App Icons

Personalizing app icons adds a layer of individuality to the home screen, allowing users to express their style and preferences.

- **Icon Customization:** Users can change the appearance of app icons using custom images or designs. This feature can be accessed through the Shortcuts app, where users create shortcuts with custom icons that launch specific apps. This allows for a cohesive and visually appealing home screen that reflects personal tastes.
- **Theme Packs:** iOS 18 supports theme packs, which are collections of icons, wallpapers, and widgets designed to work together. Users can download theme packs from the App Store or third-party developers to give their home screens a consistent and stylish look.
- **Dynamic Icons:** Some apps offer dynamic icons that change based on context or user activity. For example, a weather app icon might display current weather conditions, or a fitness app might show the user's daily step count. These dynamic icons add an interactive and informative element to the home screen.

Protecting Sensitive Information

As users customize their devices, protecting sensitive information becomes paramount. iOS 18 introduces features designed to safeguard personal data and maintain privacy.

Locked and Hidden Apps

Managing sensitive apps ensures that personal information remains secure, even if others have access to the device.

- **App Locking:** Users can lock specific apps with Face ID, Touch ID, or a passcode, preventing unauthorized access. This feature is particularly useful for apps that contain sensitive information, such as banking or messaging apps. Users can choose which apps to lock through the device settings, providing an additional layer of security.
- **Hidden Apps:** For a more discreet approach, users can hide apps from the home screen. Hidden apps remain installed on the device and can be accessed through the App Library or by using search, but they do not appear on the main home screens. This feature is useful for keeping rarely used or sensitive apps out of sight while maintaining quick access when needed.
- **Privacy Indicators:** iOS 18 includes visual indicators to show when an app is accessing sensitive information, such as the microphone or camera. These indicators provide transparency and allow users to monitor app activity, ensuring that apps only access sensitive data when necessary.

The New Control Center

The redesigned Control Center in iOS 18 offers a more streamlined and customizable interface, providing quick access to essential controls and settings.

Redesigned Interface

The Control Center's new interface is designed to be more intuitive and user-friendly, allowing users to manage settings and controls with ease.

- **Modular Layout:** The Control Center features a modular layout that allows users to add, remove, and rearrange controls according to their preferences. This modular approach enables users to tailor the Control Center to their needs, ensuring that frequently used controls are always within easy reach.
- **Quick Actions:** The redesigned interface includes quick actions for commonly used functions, such as toggling Wi-Fi, adjusting screen brightness, or controlling media playback. These quick actions are prominently displayed, making it easy to perform essential tasks with a single tap.
- **Visual Enhancements:** The Control Center has been visually enhanced with a modern design that incorporates transparency,

rounded corners, and subtle animations. These visual improvements create a cohesive and aesthetically pleasing experience that aligns with the overall look and feel of iOS 18.

Adding and Customizing Controls

Customizing the Control Center allows users to access the tools and settings they use most frequently.

- **Control Additions:** Users can add a variety of controls to the Control Center, such as shortcuts to specific settings, access to smart home devices, or controls for accessibility features. This customization ensures that the Control Center is tailored to the user's lifestyle and needs.
- **Control Organization:** Users can organize the placement of controls within the Control Center, prioritizing the most important ones for quick access. This organizational flexibility allows users to create a logical and efficient layout that enhances usability.
- **Third-Party Integrations:** iOS 18 supports third-party app integrations within the Control Center, allowing users to add controls for compatible apps. This feature extends the functionality of the Control Center, providing a unified platform for

managing both system settings and third-party app features.

The extensive customization options in iOS 18 empower users to transform their devices into personalized tools that align with their individual preferences and lifestyles. From organizing the home screen to securing sensitive information and tailoring the Control Center, iOS 18 offers a comprehensive suite of features designed to enhance the user experience. As we explore the next chapter, we will dive into the enhanced security measures of iOS 18, focusing on how Apple is leading the way in protecting user data and privacy in an increasingly connected world.

Chapter 4: Photos and Memories

The Redesigned Photos App

With iOS 18, Apple has given the Photos app its biggest redesign ever, making it more intuitive and organized than before. Here are some key enhancements:

Automatic Organization

The redesigned Photos app now uses advanced machine learning to automatically organize your entire photo library. This means that your photos and videos are categorized in a way that makes it easier to find and relive your favorite moments.

- **Collections by Topics**: Your photos are grouped into collections by topics such as Recent Days, Trips, and People & Pets. This allows you to quickly browse through your photos based on specific themes or events.
- **Pinned Collections**: iOS 18 introduces Pinned Collections, a feature that lets you pin the collections or albums that are most important to you. This gives you lightning-fast access to the photos and videos you view most frequently.

Using the Carousel

One of the standout features of the redesigned Photos app is the new Carousel. This feature provides a beautiful, poster-like view of your best content and highlights a new set of photos each day for a fun surprise.

- **Carousel View**: Swipe right from the grid view to access the Carousel. This view showcases your favorite photos in a visually appealing format, making it easy to browse through your greatest hits.
- **Daily Highlights**: Each day, the Carousel updates to display a different set of photos, giving you a new perspective on your memories and making it a delightful experience to revisit your past.

Creating and Sharing Memory Movies

iOS 18 makes it easier than ever to create and share memory movies, allowing you to compile your favorite moments into beautiful, shareable videos.

- **Automatic Memory Movies**: The Photos app can automatically create memory movies based on events, trips, or specific people and pets. These movies are generated using advanced machine learning algorithms that select the best photos and videos, add music, and create smooth transitions.

- **Customization Options**: You can customize your memory movies by choosing different themes, music, and even adding or removing specific photos and videos. This ensures that your memory movies reflect your personal style and preferences.
- **Sharing Made Easy**: Once your memory movie is ready, sharing it with friends and family is simple. You can share directly from the Photos app to social media platforms, via email, or through messaging apps. This makes it easy to share your special moments with the people who matter most.

Practical Tips and Tricks

Here are some practical tips and tricks to help you make the most out of the redesigned Photos app in iOS 18:

- **Use the Search Bar**: The improved search functionality in the Photos app allows you to find specific photos and videos quickly. You can search by date, location, or even the content of the photo, such as "beach" or "birthday."
- **Utilize Albums**: Creating albums is a great way to organize your photos for specific projects or events. You can easily add photos to albums and even create shared albums that allow others to contribute their photos.

- **Explore Memories**: The Memories feature in the Photos app curates collections of photos and videos from significant moments in your life. Take some time to explore these memories, as they often highlight forgotten moments and provide a wonderful way to reminisce.

The redesigned Photos app in iOS 18 brings a host of new features and improvements that make managing and reliving your photo library more enjoyable than ever. With automatic organization, the new Carousel view, and the ability to create and share memory movies, iOS 18 ensures that your favorite moments are always at your fingertips.

Chapter 5: Enhanced Communication

New Features in Messages

iOS 18 introduces several new features to the Messages app, enhancing the way users communicate and interact with each other. These updates make messaging more expressive, versatile, and reliable.

Text Effects and Animations

The Messages app in iOS 18 now includes a variety of new text effects and animations that make conversations more engaging and fun.

- **Text Formatting**: In addition to bold, italics, underline, and strikethrough, you can now apply playful, animated effects to any letter, word, phrase, or emoji in iMessage. These effects can be automatically suggested as you type, adding a dynamic touch to your messages.
- **Animated Effects**: There are several new animated effects available, including bubble effects and screen effects that animate the entire conversation screen. These can be triggered manually or automatically based on keywords.

Satellite Messaging

A groundbreaking feature in iOS 18 is Satellite Messaging, allowing users to stay connected even when they are without Wi-Fi or cellular service.

- **Emergency Use**: This feature is particularly useful in emergency situations where traditional communication networks are unavailable. Users can send messages via satellite to emergency services or designated contacts.
- **Limited Availability**: Satellite Messaging will initially be available in the United States on iPhone 14 and later models, integrating with existing satellite features.

Scheduling Messages

iOS 18 introduces the ability to schedule messages, making it easier to manage your communication.

- **Send Later**: Whether it's too late at night or too important to forget, you can now schedule a message to send at a specified time. This is useful for sending reminders or ensuring timely communication.
- **User Interface**: The scheduling feature is seamlessly integrated into the Messages app, allowing you to choose the date and time with a simple, intuitive interface.

RCS Support

To enhance compatibility with non-iMessage users, iOS 18 includes support for RCS (Rich Communication Services) messaging.

- **Rich Media**: RCS messages support richer media, such as high-resolution photos, videos, and audio messages, providing a more complete communication experience.
- **Delivery and Read Receipts**: RCS also brings delivery and read receipts, ensuring that users can see when their messages have been delivered and read, similar to iMessage.

Mail Enhancements

The Mail app in iOS 18 receives significant updates, making it more efficient and user-friendly.

Automatic Categorization

One of the major improvements in the Mail app is the ability to automatically categorize emails.

- **Primary Category**: The Primary category helps you focus on what matters most, such as time-sensitive messages and emails from friends, family, or colleagues.
- **Categorized Viewing**: Messages are grouped into categories like receipts, marketing emails,

and newsletters, making it easier to scan and manage your inbox.

Focused Inbox Features

The Focused Inbox in iOS 18 is designed to help you manage your emails more effectively.

- **Snippets and Previews**: View snippets of messages grouped by sender, allowing you to quickly see the most important parts of your emails without opening them.
- **Enhanced Sorting**: The Focused Inbox sorts emails based on importance and relevance, ensuring that critical messages are always at the top.

Practical Tips and Tricks

Here are some tips and tricks to help you make the most out of the new communication features in iOS 18:

- **Explore Text Effects**: Experiment with the new text effects and animations in Messages to add a personal touch to your conversations.
- **Utilize Satellite Messaging**: Familiarize yourself with Satellite Messaging for emergency situations, ensuring you know how to use it if needed.
- **Schedule Important Messages**: Take advantage of the scheduling feature to plan your

communications in advance and ensure timely delivery.
- **Organize Your Mail**: Use the automatic categorization and focused inbox features to keep your email organized and manageable.

The enhanced communication features in iOS 18 make staying in touch more engaging and efficient. With new text effects and animations, Satellite Messaging, scheduled messages, and RCS support, the Messages app is more versatile than ever. Meanwhile, the Mail app's automatic categorization and focused inbox features ensure that you can manage your email with ease.

Chapter 6: Browsing and Security

iOS 18 brings significant advancements in browsing and security, making your online experience more efficient and secure. This chapter delves into the new features in Safari, the Passwords app, and Maps and Navigation.

Safari Highlights and Redesigned Reader
Highlights for Easy Browsing

Safari in iOS 18 introduces Highlights, a feature designed to streamline your browsing experience.

- **Automatic Detection**: Highlights automatically detect relevant information on a webpage, such as directions, quick links, and key facts, presenting them in a concise, accessible manner.
- **Quick Access**: This feature saves time by allowing you to quickly access important information without scrolling through the entire page. For example, when reading an article, Highlights can pinpoint critical points, directions, or background information about mentioned people, music, movies, and TV shows.

Using the New Reader Mode

Safari's Reader mode has been redesigned to offer a more intuitive and efficient reading experience.

- **Table of Contents**: The new Reader mode now includes a table of contents, helping you navigate long articles and documents with ease.
- **High-Level Summary**: Before diving into the full content, you can view a high-level summary to get the gist of an article. This feature is particularly useful for quickly assessing whether the article meets your needs.
- **Customization Options**: Reader mode also offers more customization options, allowing you to adjust font size, style, and background color to suit your reading preferences.

The Passwords App
Centralized Credential Management

The new Passwords app in iOS 18 centralizes all your credentials, making it easier to manage your passwords and security information.

- **All in One Place**: From passwords to verifications and security alerts, the Passwords app securely stores all your credentials in one location. This centralization simplifies the process of finding and managing your login information.
- **Enhanced Security**: The app provides robust security features to protect your credentials,

including biometric authentication and encryption.

Synchronization Across Devices

iOS 18 ensures that your passwords and credentials are synchronized across all your Apple devices.

- **Seamless Syncing**: Whether you're using an iPhone, iPad, Mac, Apple Vision Pro, or even Windows, your credentials are seamlessly synchronized. This means you can access your passwords and security information from any device, ensuring you always have the information you need at your fingertips.
- **Auto-Fill Integration**: The Passwords app integrates with Safari and other apps to auto-fill your login information, making it easier and faster to log into websites and applications.

Maps and Navigation
Topographic Maps

iOS 18 enhances the Maps app with detailed topographic maps, providing more comprehensive navigation information.

- **Lay of the Land**: Topographic maps offer detailed views of the terrain, including elevation changes, natural features, and man-made structures. This information is valuable for

hikers, adventurers, and anyone needing a detailed understanding of the landscape.
- **Enhanced Navigation**: These maps provide better guidance for outdoor activities, ensuring you can navigate complex terrains with confidence.

Offline Hiking Routes

For outdoor enthusiasts, iOS 18 introduces the ability to save hiking routes for offline access.

- **Download Routes**: Browse thousands of hiking routes and save them to your device. This feature is particularly useful when you're in areas with limited or no cellular coverage.
- **Custom Routes**: You can create custom walking and hiking routes, adding your notes and waypoints. This customization allows you to plan your hikes precisely, ensuring you stay on track and enjoy your adventure.
- **Offline Navigation**: With offline access, you can use these routes without relying on an internet connection, ensuring you have access to your navigation information even in remote areas.

Practical Tips and Tricks

Here are some tips and tricks to help you make the most out of the new browsing and security features in iOS 18:

- **Utilize Safari Highlights**: Take advantage of Highlights to quickly find relevant information on webpages, saving time and enhancing your browsing experience.
- **Customize Reader Mode**: Adjust Reader mode settings to fit your reading preferences, making long articles and documents more comfortable to read.
- **Secure Your Credentials**: Use the Passwords app to centralize and secure your credentials, ensuring you have easy access while keeping your information safe.
- **Plan Your Hikes**: Make the most of topographic maps and offline hiking routes in the Maps app to plan and navigate your outdoor adventures.

The enhancements in browsing and security in iOS 18 make your online experience more efficient and secure. Safari Highlights and the redesigned Reader mode streamline browsing, while the Passwords app centralizes and secures your credentials. Additionally, the Maps app's new topographic maps and offline hiking routes provide detailed navigation information for outdoor activities.

Chapter 7: Entertainment and Gaming

iOS 18 brings exciting new features and enhancements that elevate your entertainment and gaming experience. This chapter explores the innovations in Game Mode and the latest AirPods features, designed to provide superior performance, responsiveness, and audio quality.

Game Mode
Performance Optimization

iOS 18 introduces Game Mode, a feature designed to optimize your device's performance specifically for gaming.

- **Resource Allocation**: Game Mode prioritizes CPU and GPU resources for gaming applications, ensuring smoother gameplay and faster load times. This means that other background processes are minimized to free up more resources for your games.
- **Battery Management**: The mode intelligently manages power consumption, striking a balance between performance and battery life. It ensures that you get the best gaming experience without quickly draining your battery.
- **Thermal Control**: To prevent overheating during extended gaming sessions, Game Mode

includes thermal management features that adjust performance to maintain optimal device temperature.

Enhanced Responsiveness

In addition to performance optimization, Game Mode enhances the responsiveness of your gaming experience.

- **Reduced Latency**: By minimizing input lag and touch response times, Game Mode ensures that your actions are registered instantly, providing a more immersive and responsive gaming experience.
- **Networking Prioritization**: For online multiplayer games, Game Mode prioritizes network traffic to reduce latency and improve connection stability. This results in smoother gameplay and fewer interruptions during critical moments.
- **Customizable Settings**: Users can customize Game Mode settings to match their preferences, such as adjusting the sensitivity of touch controls or enabling specific performance boosts for different types of games.

AirPods Features
Hands-Free Siri

AirPods get even smarter with hands-free Siri capabilities in iOS 18.

- **Always-On Listening**: AirPods now support always-on listening for Siri commands. This means you can activate Siri without needing to touch your AirPods or device, simply by saying "Hey Siri."
- **Enhanced Commands**: The range of commands Siri can handle has been expanded, allowing for more complex interactions. You can control your music, send messages, get directions, and much more, all without lifting a finger.
- **Improved Voice Recognition**: With enhanced voice recognition technology, Siri can better understand and respond to your commands, even in noisy environments.

Voice Isolation

iOS 18 introduces Voice Isolation, a feature that enhances the clarity of your voice during calls and voice chats.

- **Noise Reduction**: Voice Isolation uses advanced algorithms to reduce background noise, focusing on your voice. This ensures that the person on the other end hears you clearly, even in noisy surroundings.
- **Adaptive Filtering**: The feature adapts to changing environmental conditions, continuously optimizing the microphone settings to provide the best possible voice quality.
- **Enhanced Communication**: Whether you're on a phone call, using FaceTime, or participating

in a gaming voice chat, Voice Isolation ensures that your communication is clear and uninterrupted.

Spatial Audio for Gaming

Spatial Audio, a feature already loved by music and movie enthusiasts, gets a gaming-specific enhancement in iOS 18.

- **Immersive Soundscapes**: Spatial Audio creates a three-dimensional audio environment, making game sounds more realistic and immersive. You can hear sounds from all directions, enhancing your situational awareness and making games more engaging.
- **Dynamic Head Tracking**: This feature adjusts the audio based on the movement of your head, ensuring that the sound stays consistent and realistic no matter how you move. This adds an extra layer of immersion, especially in VR and AR games.
- **Game-Specific Optimizations**: Some games will have specific optimizations for Spatial Audio, providing an even more tailored audio experience. This means game developers can integrate Spatial Audio features directly into their games for enhanced effects.

iOS 18 brings significant advancements to entertainment and gaming with the introduction of Game Mode and enhanced AirPods features. Game Mode ensures optimal performance and responsiveness for a superior gaming experience, while the new AirPods features—hands-free Siri, Voice Isolation, and Spatial Audio for Gaming— provide improved audio quality and convenience.

Chapter 8: Financial Management with Wallet

iOS 18 introduces several new features and enhancements to the Wallet app, making it easier and more secure to manage your finances and transactions. This chapter will delve into the innovative Tap to Cash feature, Apple Pay enhancements, and the new capabilities for managing event tickets.

Introducing Tap to Cash

Tap to Cash is a groundbreaking feature in iOS 18 that simplifies private payments between iPhone users.

Private Payments

- **Seamless Transactions**: Tap to Cash allows users to transfer money instantly by simply bringing two iPhones together. This feature uses NFC (Near Field Communication) technology to ensure secure and rapid transactions.
- **Privacy Focused**: All Tap to Cash transactions are encrypted end-to-end, ensuring that your financial information remains private and secure. Apple does not store or have access to transaction details.

- **User-Friendly Interface**: The Wallet app guides you through the process, making it easy to send or receive money. You can also view your transaction history within the app, keeping track of all payments.
- **No Additional Fees**: Tap to Cash transactions do not incur any additional fees, making it a cost-effective solution for peer-to-peer payments.

Apple Pay Enhancements

iOS 18 brings several enhancements to Apple Pay, expanding its functionality and making it even more convenient for users.

Paying with Rewards

- **Reward Points Integration**: Apple Pay now supports paying with reward points from participating banks and loyalty programs. You can choose to use your accumulated points for purchases directly within the Wallet app.
- **Automatic Calculation**: When making a purchase, Apple Pay automatically calculates the best combination of reward points and money, optimizing your payment method to save you the most.
- **Track Rewards**: The Wallet app also provides a summary of your reward points balance and transaction history, making it easier to manage and utilize your rewards.

Installment Options

- **Flexible Payment Plans**: With iOS 18, Apple Pay introduces installment payment options for eligible purchases. This allows users to spread the cost of their purchases over multiple payments, making it easier to manage larger expenses.
- **Transparent Terms**: The Wallet app displays all relevant details, including the number of installments, interest rates (if any), and payment schedule. This transparency helps users make informed financial decisions.
- **Automatic Payments**: Users can set up automatic payments for their installments, ensuring they never miss a due date. Notifications and reminders help keep track of upcoming payments.

Managing Event Tickets

iOS 18 enhances the Wallet app's capabilities for managing event tickets, providing a more organized and user-friendly experience.

New Features and Guides

- **Redesigned Event Tickets**: Event tickets in the Wallet app have been redesigned for better visibility and functionality. Important details such as event time, venue, and seat number are prominently displayed.

- **Event Guides**: Each event ticket now includes a comprehensive event guide. This guide provides useful information about the venue, including parking options, entry points, and nearby amenities. It also offers recommendations from Apple apps, such as nearby restaurants and attractions.
- **Enhanced Notifications**: Users receive real-time updates about their events, such as changes in event time or venue, ensuring they stay informed. The Wallet app also sends reminders as the event date approaches.
- **Sharing Tickets**: Sharing event tickets with friends and family is easier than ever. You can share tickets directly from the Wallet app, allowing others to access and use the tickets seamlessly.
- **Digital Ticket Scanning**: The Wallet app supports digital ticket scanning, allowing for quick and contactless entry at events. This feature is designed to work with most major event venues and ticketing systems.

iOS 18's enhancements to the Wallet app make it a powerful tool for managing your financial transactions and event tickets. Tap to Cash offers a secure and convenient way to make private payments, while Apple Pay's new features allow for flexible and reward-based payments. The improvements in managing event tickets provide a seamless experience for attending and enjoying events.

These innovations in financial management and ticketing underscore Apple's commitment to providing a secure, user-friendly, and comprehensive digital wallet. With these new capabilities, iOS 18 empowers users to handle their finances and event plans with greater ease and confidence.

Chapter 9: Productivity and Well-being

iOS 18 is designed to enhance both productivity and well-being through innovative features and apps. This chapter explores the enhancements to the Notes app and the introduction of the Journal app, both of which aim to improve how users manage their tasks, track their well-being, and achieve their goals.

Enhanced Notes App

The Notes app in iOS 18 has been significantly upgraded to include features that boost productivity and make note-taking more powerful and versatile.

Live Audio Transcription

- **Real-Time Transcription**: The Notes app now offers live audio transcription, allowing users to convert spoken words into text in real time. This feature is particularly useful for capturing meeting notes, lectures, and brainstorming sessions.
- **Accurate and Fast**: Utilizing advanced speech recognition technology, the transcription is highly accurate and quickly processes spoken words, ensuring minimal lag between speech and text display.

- **Editable Transcriptions**: Users can easily edit transcriptions to correct any errors or add additional notes. The transcriptions are saved within the note, making it easy to reference and review later.
- **Searchable Text**: Transcribed text is fully searchable within the Notes app, allowing users to quickly find specific information or keywords within their notes.

Crunching Numbers in Notes

- **Embedded Calculations**: iOS 18 introduces the ability to perform calculations directly within the Notes app. Users can type mathematical expressions, and the app will automatically calculate and display the results.
- **Spreadsheet Integration**: For more complex calculations, Notes now supports basic spreadsheet functions. Users can create tables, input data, and perform operations similar to those found in dedicated spreadsheet applications.
- **Graphical Data Representation**: Notes can also generate simple charts and graphs based on the data entered, providing a visual representation of numerical information. This feature is useful for creating budgets, tracking expenses, or visualizing data trends.

The Journal App

The new Journal app in iOS 18 is designed to help users log their well-being, track goals, and integrate with the Health app to provide a comprehensive overview of their mental and physical health.

Logging Well-being

- **Daily Logs**: The Journal app allows users to log their daily activities, thoughts, and feelings. This helps in tracking mood patterns and identifying factors that influence well-being.
- **Prompted Entries**: Users receive prompts and suggestions for journal entries, encouraging regular use and helping to reflect on various aspects of their life. Prompts can be customized based on user preferences.
- **Multimedia Support**: Journal entries can include photos, videos, and audio recordings, providing a richer and more detailed account of daily experiences.

Tracking Goals and Insights

- **Goal Setting**: The Journal app enables users to set personal goals and track their progress over time. Goals can be related to health, fitness, personal development, or any other area of interest.
- **Progress Tracking**: Users can view their progress through visual indicators such as charts

and graphs, helping them stay motivated and on track. The app provides reminders and notifications to keep users engaged with their goals.

- **Insight Generation**: The Journal app analyzes entries and provides insights into patterns and trends. For example, it might identify correlations between certain activities and mood changes, helping users understand what positively or negatively affects their well-being.

Integrating with the Health App

- **Seamless Integration**: The Journal app integrates seamlessly with the Health app, allowing users to sync health data such as activity levels, sleep patterns, and vital signs. This integration provides a holistic view of both mental and physical health.
- **Health Metrics**: Users can set goals based on health metrics tracked by the Health app, such as steps taken, calories burned, or hours slept. The Journal app can suggest activities and tips to improve health based on this data.
- **Comprehensive Reports**: Combined data from the Journal and Health apps can be compiled into comprehensive reports, giving users detailed insights into their overall well-being. These reports can be shared with healthcare providers if desired.

iOS 18's enhancements to the Notes app and the introduction of the Journal app reflect Apple's commitment to improving productivity and well-being. The Notes app's new features, such as live audio transcription and embedded calculations, make it a powerful tool for capturing and organizing information. The Journal app provides users with a comprehensive platform to log their well-being, track goals, and integrate with the Health app, promoting a holistic approach to mental and physical health.

Chapter 10: Privacy, Security, and Accessibility

In iOS 18, Apple continues to prioritize user privacy, security, and accessibility, introducing advanced features and enhancements that empower users and protect their personal information. This chapter explores the latest innovations in privacy protection, accessibility improvements, and other key updates designed to make iOS 18 more secure and inclusive than ever before.

Advanced Privacy Features

Privacy is at the forefront of iOS 18, with new features that give users more control over their personal information and enhance security across the platform.

Contact Sharing Control

iOS 18 introduces enhanced controls for sharing contacts with apps, providing users with greater transparency and control over their contact information.

- **Selective Sharing:** Users can now choose which contacts to share with specific apps, rather than granting access to all contacts by default. This selective sharing feature ensures that apps only receive contact information when explicitly authorized by the user, enhancing privacy and minimizing data exposure.
- **Permission Management:** The updated privacy settings menu in iOS 18 includes improved management options for contact sharing permissions. Users can review and modify which apps have access to their contacts at any time, empowering them to maintain control over their personal data.

Improved Bluetooth Privacy

iOS 18 enhances Bluetooth privacy protections, safeguarding user devices from potential tracking and unauthorized access.

- **Address Randomization:** iOS 18 includes enhanced address randomization techniques for Bluetooth connections, making it more difficult for malicious actors to track or identify devices based on Bluetooth signals. This feature strengthens privacy protections while maintaining seamless connectivity for legitimate device interactions.

- **Permission Prompts:** When apps request Bluetooth access, iOS 18 provides clear permission prompts that explain why access is needed and allow users to grant or deny permission accordingly. This transparency ensures that Bluetooth functionality is used responsibly and in accordance with user preferences.

Accessibility Innovations

iOS 18 introduces groundbreaking accessibility features that make iPhones more accessible and inclusive for users with diverse needs.

Eye Tracking for Control

Eye tracking technology enables users to control their iPhones using just their eyes, providing a revolutionary method of interaction for individuals with mobility impairments.

- **Hands-Free Navigation:** With eye tracking enabled, users can navigate the iPhone interface, select apps, type messages, and perform other tasks using intuitive eye movements. This hands-free control enhances independence and accessibility, empowering users to interact with their devices more effectively.

- **Customizable Settings:** iOS 18 includes customizable settings for eye tracking sensitivity and calibration, allowing users to personalize their eye control experience based on their individual preferences and needs. These settings ensure optimal performance and comfort during device use.

Music Haptics for the Hearing Impaired

Music Haptics is a groundbreaking feature in iOS 18 that enhances the music listening experience for users who are deaf or hard of hearing.

- **Taptic Engine Integration:** iOS 18 syncs the Taptic Engine with the rhythm and beats of music playback, translating auditory elements into tactile vibrations. This tactile feedback allows users to perceive the tempo, intensity, and emotional nuances of music through vibration patterns, providing a richer and more immersive listening experience.
- **Customization Options:** Users can adjust Music Haptics settings to fine-tune vibration intensity, pattern dynamics, and synchronization with music playback. These customization options cater to individual preferences and sensory sensitivities, ensuring a personalized and enjoyable music experience for all users.

Other Key Improvements

iOS 18 introduces several additional enhancements that improve functionality, productivity, and user experience across various apps and features.

Calculator and Calendar Enhancements

- **Math Notes Calculator:** iOS 18 includes the all-new Math Notes calculator, which supports advanced mathematical functions, unit conversions, and live calculations. This feature is designed to simplify complex calculations and enhance productivity for users who rely on mathematical tools in their daily tasks.
- **Redesigned Calendar Month View:** The Calendar app in iOS 18 features a redesigned month view, providing users with a clearer and more intuitive overview of their upcoming events and appointments. Enhanced navigation and visual clarity make it easier to manage schedules and plan ahead effectively.

Freeform and SharePlay Updates

- **Freeform Scenes:** iOS 18 enhances Freeform with new organizational tools, including section-by-section organization and improved diagramming capabilities. These updates streamline content creation and collaboration, enabling users to create structured and visually appealing presentations with ease.

- **Enhanced SharePlay:** SharePlay in iOS 18 offers enhanced screen sharing capabilities, allowing users to draw on shared screens and control remote devices during collaborative sessions. These interactive features facilitate seamless communication and collaboration across devices, enhancing productivity and engagement.

Emergency SOS with Live Video

- **Live Video Sharing:** iOS 18 introduces Emergency SOS with live video sharing capabilities, enabling users to share streaming video and recorded media during emergency calls. This feature facilitates faster and more effective emergency response by providing real-time visual information to emergency dispatchers, enhancing safety and peace of mind for users in critical situations.

iOS 18 sets new benchmarks in privacy protection, security enhancements, and accessibility innovations, reaffirming Apple's commitment to empowering users and fostering inclusivity. By introducing advanced privacy controls, groundbreaking accessibility features like eye tracking and Music Haptics, and key improvements across core apps and functionalities, iOS 18 delivers a comprehensive and user-centric experience. As we look ahead, the next chapter will explore the evolving landscape of digital health features in iOS 18, highlighting how Apple is leveraging technology to promote well-being and empower users to live healthier lives.

Chapter 11: The Future of iOS and Apple's Vision

In Chapter 11, we delve into the future of iOS and Apple's overarching vision, exploring upcoming technologies, strategic directions, and practical advice for users to maximize their Apple devices.

VisionOS 2 and Spatial Computing

Apple's commitment to advancing spatial computing through VisionOS 2 marks a significant leap forward in immersive technology.

Spatial Delivery

VisionOS 2 introduces enhanced spatial capabilities, leveraging advanced sensors and AI to create a more intuitive and immersive computing experience. From enhanced AR applications to seamless integration with physical spaces, VisionOS 2 transforms how users interact with digital content.

- **AR Enhancements:** VisionOS 2 enhances augmented reality experiences with improved object recognition, spatial audio, and real-time rendering. Users can expect more immersive AR apps that blend digital content seamlessly into their physical surroundings.

- **Spatial Navigation:** The integration of spatial computing in VisionOS 2 allows for precise indoor navigation, virtual mapping, and interactive experiences that enhance productivity and engagement in various contexts.

The Integration of AI and Machine Learning

iOS's evolution into a smarter, more intuitive platform is driven by Apple's robust integration of AI and machine learning technologies.

Apple Intelligence

- **Personal Context Awareness:** Apple Intelligence in iOS 18 harnesses personal data and user behavior to provide tailored insights and recommendations. This contextual awareness enhances user experience across apps, from predictive text suggestions to adaptive app recommendations.
- **AI-Driven Enhancements:** Machine learning algorithms power intelligent features like Siri's improved language

understanding and predictive capabilities. These advancements enable Siri to anticipate user needs, streamline daily tasks, **and deliver more personalized interactions.**

Preparing for Future Updates

As Apple continues to innovate, users must stay informed and prepared for upcoming iOS updates.

Software Update Strategies

- **Benefits of Regular Updates:** Stay ahead with enhanced security features, performance optimizations, and access to new functionalities introduced in each iOS update.
- **Update Preparation Tips:** Ensure device compatibility, backup essential data, and familiarize yourself with new features before installing updates to maximize benefits and minimize disruption.

How to Stay Informed and Get the Most Out of Your Apple Devices

Empower yourself with knowledge and practical tips to optimize your Apple devices and leverage new technologies effectively.

Continuous Learning and Support

- **Official Resources:** Access Apple's official support channels, user guides, and online forums for troubleshooting tips, feature explanations, and community support.
- **Educational Content:** Explore Apple's educational initiatives, including Today at Apple sessions and online tutorials, to expand your knowledge and skills in using iOS devices.
- **Community Engagement:** Join online communities, attend local Apple Store events, and participate in user groups to share experiences, learn from others, and stay updated on the latest iOS developments.

Chapter 11 concludes with a forward-looking perspective on the future of iOS and Apple's vision for technological innovation. From VisionOS 2's advancements in spatial computing to the transformative integration of AI and machine learning, iOS continues to evolve as a powerful platform that enriches user experiences and drives digital innovation. By preparing for future updates and equipping yourself with knowledge on optimizing Apple devices, you can harness the full potential of iOS and stay at the forefront of technology. Embrace the future with confidence, knowing that Apple remains committed to delivering cutting-edge solutions that redefine how we interact with technology and each other.

Conclusion

As we conclude our exploration of iOS 18, it becomes evident that Apple has once again set new standards for innovation, user experience, and technological advancement. Throughout this book, we have delved into the myriad features and enhancements that make iOS 18 a pivotal update in the evolution of mobile operating systems. Let's recap the advantages of iOS 18, encourage adaptation and exploration, and reflect on the future role of technology in our daily lives.

Summarizing the Advantages of iOS 18

iOS 18 introduces a wealth of features designed to enhance productivity, creativity, and personalization:

- **Enhanced Privacy and Security:** Advanced controls for contact sharing and improved Bluetooth privacy settings ensure that your personal information remains protected.
- **Customization and Personalization:** From redesigned Home Screen layouts to personalized widgets and app icons, iOS 18 empowers users to tailor their devices to reflect their unique preferences and workflows.

- **Intelligent Features:** Apple Intelligence and AI-driven enhancements in Siri and other apps leverage personal context to provide tailored recommendations and streamline everyday tasks.
- **Accessibility Innovations:** Innovations such as eye tracking for control and Music Haptics for the hearing impaired underscore Apple's commitment to inclusivity and accessibility.

Encouraging Adaptation and Exploration

Embracing iOS 18 means embracing a future where technology enriches our lives in profound ways:

- **Seamless Integration:** Explore how iOS 18 seamlessly integrates with other Apple devices and services, enhancing connectivity and continuity across your digital ecosystem.
- **Creative Expression:** Utilize new tools for image creation, Genmoji crafting, and memory movies to express yourself creatively and relive memorable moments in innovative ways.
- **Productivity and Efficiency:** Discover how iOS 18's redesigned Control Center, improved writing tools, and organizational features like Collections and Carousel can boost your productivity and simplify daily tasks.

As technology continues to evolve, iOS 18 serves as a testament to the transformative power of innovation:

- **Digital Well-being:** Consider how iOS 18's features, such as Journal for well-being and personalized fitness insights, contribute to a balanced and mindful approach to digital engagement.
- **Future Trends:** Anticipate upcoming trends in mobile technology, including advancements in augmented reality, spatial computing, and the integration of AI in everyday applications.
- **Empowering Users:** Reflect on how iOS 18 empowers users to harness the full potential of their devices, fostering creativity, productivity, and connectivity in an increasingly digital world.

Final Thoughts

As we navigate the ever-evolving landscape of technology, iOS 18 stands as a testament to Apple's commitment to innovation, user-centric design, and pushing the boundaries of what is possible. Whether you are a longtime Apple enthusiast or a new user exploring the iOS ecosystem, embracing iOS 18 promises an enriching and transformative experience. Let us continue to embrace the future with curiosity, adaptability, and a commitment to leveraging technology for positive change in our lives and communities. Together, we shape the future of digital innovation, guided by the ethos of creativity, inclusivity, and empowerment that defines iOS 18 and the Apple ecosystem.

www.ingramcontent.com/pod-product-compliance
Lightning Source LLC
Chambersburg PA
CBHW071952210526
45479CB00003B/907